DATE DUE

APR 0 4		
NOV 2 1		
MAY 0 8 2000		
AUG 2 6 2000		

HOW DID WE FIND OUT
ABOUT COAL?

How Did We Find Out . . . ?
Books by Isaac Asimov

HOW DID WE FIND OUT THE EARTH IS ROUND?

HOW DID WE FIND OUT ABOUT ELECTRICITY?

HOW DID WE FIND OUT ABOUT NUMBERS?

HOW DID WE FIND OUT ABOUT DINOSAURS?

HOW DID WE FIND OUT ABOUT GERMS?

HOW DID WE FIND OUT ABOUT VITAMINS?

HOW DID WE FIND OUT ABOUT COMETS?

HOW DID WE FIND OUT ABOUT ENERGY?

HOW DID WE FIND OUT ABOUT ATOMS?

HOW DID WE FIND OUT ABOUT NUCLEAR POWER?

HOW DID WE FIND OUT ABOUT OUTER SPACE?

HOW DID WE FIND OUT ABOUT EARTHQUAKES?

HOW DID WE FIND OUT ABOUT BLACK HOLES?

HOW DID WE FIND OUT ABOUT OUR HUMAN ROOTS?

HOW DID WE FIND OUT ABOUT ANTARCTICA?

HOW DID WE FIND OUT ABOUT OIL?

HOW DID WE FIND OUT ABOUT COAL?

HOW DID WE FIND OUT

ABOUT COAL?

Isaac Asimov
Illustrated by David Wool

WALKER AND COMPANY
New York

Library of Congress Cataloging in Publication Data

Asimov, Isaac, 1920–
  How did we find out about coal?

  (How did we find out . . . series)
  Includes index.
  SUMMARY: Presents a history of fire, considers
wood as a fuel, and discusses the formation of
coal and the history of its use as a fuel.
  1. Coal—Juvenile literature. [1. Coal]
I. Wool, David. II. Title.
TP325.A78   1980   553.2′4   80-50448
ISBN 0-8027-6400-2
ISBN 0-8027-6401-0 (lib. bdg.)

First published in the United States of America
in 1980 by the Walker Publishing Company, Inc.

Published simultaneously in Canada by Beaverbooks,
Limited, Pickering, Ontario.

Trade ISBN: 0-8027-6400-2
Reinf. ISBN: 0-8027-6401-0

Library of Congress Catalog Card Number: 80-50448

Printed in the United States of America

10 9 8 7 6 5 4 3 2 1

To
Shawna McCarthy,
a red-haired diamond.

HOW DID WE FIND OUT ... SERIES
Each of the books in this series on the history of
science emphasizes the process of discovery.

Contents

1   Fire

EVERYONE HAS seen a fire at some time or other. We know those dancing yellow flames that give off light and heat. We see it when something is burning. We see it when wood burns, or paper, or anything else that is *inflammable.*

What makes something inflammable?

Everything is made up of tiny atoms, far too tiny to be seen even in a microscope. These come in a hundred or so different varieties. Two common varieties are *carbon atoms* and *hydrogen atoms.*

Carbon atoms can combine with another atom variety called *oxygen.* Oxygen exists in the air, and when it combines with carbon, heat is produced. Hydrogen can also combine with oxygen to produce heat. It is this combination of atoms to form heat (and usually light also) that we call burning.

Inflammable substances such as wood and paper include a large number of carbon and hydrogen atoms in their makeup. These atoms, together with others, clump together in groups called *molecules.*

11

The molecules in wood and paper are large groups of atoms. These molecules make up solid substances that do not combine with oxygen when they are cool. If the wood or paper is heated, however, the large molecules are broken up into small ones that turn into hot gases, or vapors. The carbon and hydrogen atoms in the vapors combine with the oxygen in the air, producing heat and light.

Fire consists of these vapors giving off heat and light as they combine with oxygen.

Once the vapors burn and produce heat, that heat will cause other inflammable objects to burn if those other objects are close enough to be heated by the fire. If one end of a piece of paper is burning, the heat it produces will make nearby parts of the paper burn and that will make more parts burn and so on.

You can begin with a single piece of paper on fire and burn tons of paper if you keep adding it to the fire. A tiny bit of fire on a single leaf can spread and spread and burn down a mighty forest.

That sounds very dangerous, and it *is* dangerous. People must be very careful of fires at all times.

Fortunately, fires don't start easily. The first bit of fire only starts if the inflammable material is heated to a high temperature. It isn't easy to get that high temperature without a fire to begin with.

How did the first fire begin? Did a human start it?

No, there were fires on earth for long ages before human beings even existed. Once plant life

HOMO ERECTUS CHILD WITH BURNING TWIG

covered the dry land, beginning about 400 million years ago, there was always the chance of fire.

Plants are made in large part of woody material, and they are inflammable, especially when they're particularly dry because it hasn't rained for a while. Once the clouds do come, though, they are sometimes accompanied by lightning.

Lightning produces light and heat as a result of the flow of particles called electrons that are even smaller than atoms. When lightning strikes a tree, its heat can set the tree on fire. The fire can

LIGHTNING STORM

spread to other trees, and soon there is a "forest fire." This will burn till the fire reaches places where there are no other trees near enough to catch fire, or until a rain comes that is heavy enough to put it out.

Animals, if caught in a forest fire, will also burn and will die. Most animals quickly learn to fear fire and to run from it. Primitive human beings called *hominids* (see *How Did We Find Out About Our Human Roots?* Walker, 1979), who lived a million years or so ago, also feared fire and also ran away from it.

Hominids were smarter than other animals, however, and also more curious. (The two go together.)

About half a million years ago, the brainiest kind of hominid that lived was called *Homo erectus*. *Homo erectus* was not as brainy as human beings are today. (Modern human beings are *Homo sapiens*.) Still *Homo erectus* was brainier than any other land animals.

Homo erectus was so intelligent that its curiosity about fire was stronger than its fear.

After a forest fire is over, there may still be some burning pieces of twigs or branches scattered on the ground. Perhaps some *Homo erectus* children (children are even more curious than adults, of course) crept close and watched the twigs burn. They may have seen another twig catch fire. After a while, some particularly bold child might have picked up a twig that wasn't burning and placed it in the fire. Then *it* would begin to burn.

It may have been a kind of plaything at first,

and a rather dangerous one. Still, it may have occurred to some of the *Homo erectus* adults who saw what the children were doing that a fire could be good to have around if it stayed small.

Suppose you put only a small amount of inflammable material (or *fuel*) into a fire at any one time. Suppose you kept all other inflammable material a distance away from it. Then the fire would stay small. It would not spread and become large and dangerous.

A small, tame fire would give light and warmth. Other animals, even large and dangerous ones, were afraid of fire and would stay away from one. Hominids sleeping about a campfire would be safer from prowling animals than they would be if there were no fire.

All this isn't just guessing. In caves in north China, bones of *Homo erectus* were discovered that were half a million years old. And there were traces of campfires near them.

Only *Homo erectus* and the even brainier *Homo sapiens* that followed have ever tamed fire. All human beings of every kind have known how to use fire for thousands upon thousands of years. No animals of any other kind, not even the brightest, have ever known how.

As time went on, a great many further uses of fire were discovered.

For instance, it was found (perhaps by accident to begin with) that meat heated over a fire was easier to chew. It also tasted better. Such cooked food was safer to eat, too. Though primitive human beings didn't know it, the heat killed germs and other parasites in the food.

In still later ages it was found that fire could bake soft clay into hard pottery. Fire could melt sand mixed with other minerals to make glass. Fire could heat certain rocks called *ores* to produce such metals as copper, tin, and iron.

Of course, fire also had its dangers. It could spread accidentally. It could burn houses, food supplies, even people. Even when it didn't spread, it still produced smoke, which made things smelly and dirty, and which made people cough. It also left behind ashes that got in the way.

The uses of fire were far more important than the discomforts, however. People kept using fire and tried to be as careful as possible to keep it from spreading. When they kept fire in a house, they learned to build chimneys to carry off most of the smoke. They learned to collect the ashes and dump them some distance away.

One problem with a fire was just the opposite of its spreading. A fire could go out.

Every family must have worked hard to keep that from happening. One of the tasks of young children in a family might have been to collect branches, twigs, and brush to keep the fire going. Sometimes a second fire might be started by carrying a burning twig to a new pile of fuel. Then the first one could be allowed to go out and the ashes could be cleaned away.

Still, a fire might go out by accident. In that case someone might be sent to another house or even to a distant village to borrow a light from a fire there. Some twigs could be set to burning, then placed in a pot and brought back home where they could be used to start a new fire.

But what if someone's fire went out and there was no other fire within reach? What can anyone do then? Wait for lightning and for another forest fire?

The use of fire was never really satisfactory until some way was discovered of starting a fire *without* lightning and *without* another fire that was already burning. It may not have been until nine thousand years ago that people learned how to do that.

It may have happened by accident. Human beings made tools out of rocks. To shape the tools they would hit one rock with another, knocking chips off. The rubbing (or *friction*) of one rock against another heated the rocks. Sometimes the tiny fragments that were knocked off were heated till they were hot enough to glow and form sparks.

If these sparks fell on something that was inflammable, they might start a fire. Eventually, people may have learned to hit rocks together deliberately in such a way as to allow sparks to fall on dry, powdered plant material (*tinder*) and set it on fire. Then they would have a fire where there had been no fire to begin with.

Another way would be to grind a pointed stick into a hole in another stick. The friction would heat up both sticks, and if there was tinder in the hole, that would eventually catch fire.

Neither way was exactly easy, but fire was important enough to take a lot of trouble over.

We have made the system easier in modern time. In cigarette lighters, a metal wheel rubs against a kind of rock called flint. This shoots out sparks which sets inflammable vapors on fire.

EARLY MAN STARTING A FIRE

We also use the system of rubbing wood to set it on fire by friction. Nowadays, though, we coat the piece of wood with a chemical that catches fire very easily when it is heated. That give us a *match*.

Just the same, the easiest way is still to borrow fire from one that already exists. That is why we have *pilot lights* on stoves. These are small flames fed by flows of gas. When we turn on the gas burners, the gas that comes out catches fire from the pilot light.

2　Wood

ONCE PEOPLE tamed fire and had ways of starting it if it went out, there remained the problem of fuel.

The best fuel for human beings, to begin with, was wood. For one thing, except in deserts and in polar regions, wood is very common. Once it is dry, it will burn easily and not too rapidly. It burns with a fire that gives off considerable light and heat. What's more, human beings don't eat wood, so they don't have to choose between feeding the fire and feeding themselves.

When a large pile of wood is burning, the carbon and hydrogen atoms in the wood on the outside of the pile combine easily with the oxygen in the air. It is hard, though, for the air to get to the center of the pile of wood.

The center of the pile of wood is heated up and its molecules break up into vapors, but there isn't much oxygen about. What little oxygen that does manage to get into the center of the pile combines with the hydrogen atoms. The hydrogen atoms

WOOD FIRE

combine more easily with oxygen than the carbon atoms do.

At the center of the pile of wood, then, a collection of material builds up that is mostly carbon atoms. Carbon atoms when they exist by themselves make up a blackish material. You can tell it isn't ash because ash is usually white.

You can see this happening even in a burning stick. If the fire is blown out, the stick is black where the fire has been burning. That is because the hydrogen atoms combined with oxygen first and left the carbon behind. Such a stick is said to be *charred*.

All that black material at the center of a wood fire after the fire has gone out is charred wood.

The carbon atoms in the charred wood will burn if the material is placed in a fire. The carbon atoms will then finally have a chance to combine with oxygen. Carbon atoms by themselves do not give off vapors, though, so there are no dancing flames. The pieces of charred wood just grow red hot and slowly turn to ash. The dim glow of a

CHARRED WOOD

slowly burning object is called a *coal*. Because the charred material burns like that, it was called *charcoal*.

Charcoal has some advantages over wood. Charcoal burns more slowly than wood and it also burns more hotly. This makes it more useful than wood in some kinds of cooking.

Then, too, charcoal is made up almost entirely of carbon atoms and these carbon atoms can combine with oxygen that is already combined with metals.

Ores are made up of metal-oxygen combinations. When the carbon takes the oxygen out of the ore and combines with it, pure metal is left behind. The high temperature produced when charcoal burns helps this change along. Charcoal turned out to be very important in obtaining metals from ores–particularly iron.

Of course, since charcoal doesn't produce vapors, it doesn't give much light. If you want to see something at night, charcoal will never do. You would still have to have a wood fire.

The advantages of charcoal were so great that people began to make it deliberately. They would set a large wood fire burning and then cover it up loosely with soil so as to cut down the amount of oxygen that could reach it.

Naturally, a great deal of wood had to be burned away in order to produce charcoal. You had to burn several pounds of wood in order to get one pound of charcoal.

This didn't seem bothersome to human beings in early times. There were so many trees everywhere that it didn't matter how much wood was burned. There was always plenty more.

Some types of wood gave more light than others did. There were the kinds that contained soft, gummy substances called *resin*. Such wood burned with a brighter flame that made it possible to see at night. The wood of certain evergreen trees such as pines and cedars burned brightly for this reason and such wood was used as torches.

Then, too, there were other inflammable substances besides wood. Oils could be obtained from certain plants and animals, and these were inflammable liquids. Chunks of wood could be soaked in oil, and they would then burn with a brighter flame.

Or else the oil could be used by itself. A pool of oil could be placed in a hollowed-out rock, or in a clay pot, and a piece of porous material (a *wick*) could be placed in it. The oil soaked into the wick, which stuck up out of the oil. The upper end was lighted. The oil in that end slowly burned and, as it burned away, more oil soaked its way up the wick and burned in its turn.

This container of burning oil is called a *lamp*. Some very simple forms of lamps may have existed as long as 70,000 years ago.

A lamp is more convenient than a wood fire in some ways. You can carry a lamp from place to place. You can put it wherever you need it, so you can see to do your work or to read by. You can't carry wood fires about that way.

Of course, a lamp might accidentally be spilled and the burning oil might start a bad fire.

But there are also solid oils, call *fats*, which are inflammable. In addition there are *waxes*, such as that produced in beehives.

With fats or waxes, you don't need a lamp. The solid material can be heated gently till it melts. The melted fat or wax can be allowed to coat the wick. Then it is cooled so that the coating becomes solid. A thicker and thicker coating can be built up, and in the end there is a thick pipe of fat or wax with the wick running down its center. The result is a *candle*. These first appeared about 5,000 years ago.

A candle is even easier to carry about than a lamp and it can't be spilled.

However, though waxes, fats, and oils are all useful, they are not nearly as common as wood. If a large fire were needed in early times, no one could possibly expect to find a large heap of wax or fat, or gallon upon gallon of oil. It would take too much time and effort to collect all the bee-hives, or to squeeze out all the olives, or to melt down all the fat in chickens or cattle.

In the time it took to do that, people could chop down any number of trees and split them up into firewood.

Therefore, right down into modern times, wood was the chief fuel of the fires used by human beings. In many parts of the world wood is still the chief fuel. Even in the United States, people in the country (and sometime in towns, too) burn wood in fireplaces or in stoves.

Wood is not just a wonderful fuel. It is a remarkable substance in many other ways. It is strong; it lasts long; it has a beautiful appearance; it can be cut into any shape; it can be smoothed and waxed.

For that reason it can be used to build houses or ships or to make furniture or a million and one other things.

In modern times it was found that wood was the cheapest possible source out of which paper could be made and paper has a million uses, too. The paper I am typing on and the paper in the book you are reading was once part of a tree.

In many, many ways human beings depended on wood through long ages.

FOREST—A GOOD WOOD SUPPLY

3 Coal

THE USE of fire (and of all the other advances that fire made possible) made life more comfortable. The population grew larger because more and more people who were born could be kept alive.

The more people there were, the more wood was needed to make fires, build houses and ships, and make furniture. Slowly, through the years and centuries, more and more wood was used.

People didn't worry about that. There seemed to be an endless supply of trees and more trees were always growing. The forests of the world must have seemed like those magic pitchers we read about in fairy tales. No matter how much milk is poured out of such pitchers, there is always more left.

But real life is not like a fairy tale. Trees only grow so fast. Every year only so much new wood is formed. Eventually, as more and more human beings used more and more wood, the point was reached where more wood was used each year than the amount of new wood that grew.

When that point was reached, the forests began to disappear. Wood began to be more scarce. In the places where civilization had existed for centuries, wood became so scarce that it had to be imported from other places.

That made wood all the more difficult to get and all the more expensive. Many people must have begun to wish there was some other fuel that could be used that would be more common and cheaper than wood was.

Actually, another such fuel did exist. It was a fuel that was, in some ways, very much like wood. In fact, it was a fuel that had once been wood and that had existed in forests long, long ago.

These very ancient forests were not composed of modern trees but of ancient varieties that no longer exist. They were composed of plants called *horsetails, clubmosses, giant ferns*, and so on.

Beginning about 345 million years ago and continuing for over 100 million years, huge forests of these trees grew in large areas of low, flat, swampy land.

Naturally, trees only live so long, and then they die. Trees sometimes die as a result of being knocked down by lightning or by windstorms or by large animals. They can burn in forest fires, or they can gradually stop living because of old age. In all such cases, if air can get at them, the carbon and hydrogen atoms in them combine very slowly with oxygen. Eventually the trees completely decay.

Those trees that grow in swampy land, however, fall into shallow water or into bogs and mud. That makes total decay very difficult. Some decay

CARBONIFEROUS (COAL-FORMING) FOREST

31

does take place, but there is a shortage of oxygen because the open air cannot get at the fallen tree.

The same thing happened to such fallen trees as happened to burning wood when there was an oxygen shortage. The hydrogen atoms would combine with oxygen, but the carbon atoms would be left behind.

The fallen trees would slowly char, in other words, and a black material, which looked and behaved something like charcoal, would form. As more and more trees fell during hundreds, and thousands, and millions of years, the amount of black material that formed increased. Tons of it were formed; thousands of tons; millions of tons.

Once most of the forest was down, the black material was covered by mud in thicker and thicker layers. New forests grew on the mud. Then another pile of black material would form and again be covered by mud.

As the mud gets thicker and is buried deeper, its own weight squeezes the water out of it. The pieces of sand and grit in the mud stick together to form stone. The weight of the stone squeezes the black material together.

Ordinary charcoal, made by human beings, is rather light and crumbly. The black material formed from decaying trees is squeezed together so tightly that it gets heavy and hard and solid. It doesn't seem quite like charcoal. It still burns and smolders, however, so it is a kind of coal. In fact, people eventually called it *coal*.

Even today, coal is forming. There are swampy, boggy areas, where decaying plant material can be dug up and dried out to be used as fuel. This dried out material is called *peat*.

FERN LEAF IMPRINT IN COAL

Some of the hydrogen has already been lost, so peat has more carbon than fresh wood has. Fresh wood is about 50-percent carbon, but peat is about 60-percent carbon.

The next stage is *lignite*, which, when it is dry, is nearly 70-percent carbon.

Beyond that is a kind of coal which is about 85-percent carbon. If this coal is heated in the absence of air so that it doesn't burn, the 15 percent that is not carbon is driven off, along with some of the carbon. The material driven off is a black tar, or pitch, that in ancient times was called *bitumen*. This is why this kind of coal is called *bituminous coal*.

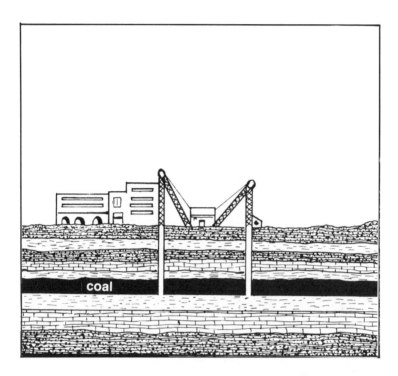

SHAFT MINE

Finally, there is a kind of coal that is at least 95-percent carbon. This burns with a red-hot glow, forming an ember, as charcoal does. The Greek word for ember is *anthrax*, so this kind of coal is called *anthracite coal*.

Coal is always formed very slowly, and it is formed much more slowly these days than in past ages, when those large forests in swampy land existed. Peat and lignite therefore make up only a small percentage of all the coal in the world. Anthracite coal only forms in a few areas where there

34

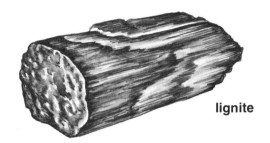

lignite

bituminous coal

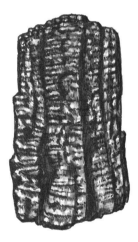

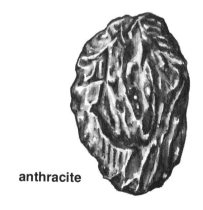

anthracite

FORMS OF COAL

35

was a great deal of pressure. It, too, makes up only a small percentage of all the coal in the world.

Most coal is bituminous coal, and there is a great deal of that under the ground. There may be as much as 8,000,000,000,000 (8 trillion) tons here and there in the earth.

As long as the coal is under the ground, people aren't likely to know that it is there. However, the ground doesn't remain unchanging during periods of millions of years.

Very slowly, the rocks in the ground under our feet change their positions. This doesn't happen while we watch. We can't *see* it happen, but it does happen—over long ages.

Layers of rock are pushed together or pulled apart as various changes take place in the earth's outermost layers. Portions of the ground are pushed upward to form mountains; other portions sink downward.

The layers of coal under the ground also heave and bend and, after millions of years, some end up farther underground than they were to begin with. Other layers end up nearer the surface. Some layers even end up at the surface so that chunks of coal can be found lying about here and there.

For thousands upon thousands of years, however, people paid no attention to the chunks of coal. They just looked like pieces of black stone. Children might have picked them up and played with them as they might with any other stone, and that was all. The black stone wasn't of a kind that was suitable for making tools, so adults were not interested.

4 The Industrial Revolution

ONE TROUBLE with coal is that it isn't easy to tell that it will burn. A material can be inflammable and yet be hard to set on fire. Whether a particular material is hard or easy to set on fire depends on its condition.

The more air that can touch an inflammable material in many different places, the more easily it can be made to start burning. A solid chunk of wood is hard to start burning, but if you take that solid chunk and split it into thin sticks, air will get at the surface of all those sticks. The sticks will be much easier to set burning than the original chunk. Sawdust will start burning still more easily.

The hydrogen atoms of the wood are more easily set on fire than the carbon atoms. The larger the percentage of carbon in a fuel, the harder it is to get it to burn. Once it is on fire, of course, it will keep on burning.

Since charcoal is almost all carbon, it is harder to start it burning than to start a piece of wood burning. Have you ever watched someone starting a charcoal fire in the backyard to grill steak or hamburgers? You may see that he burns paper first to get the charcoal started. Or else he puts an inflammable liquid on it first to get it started.

At least charcoal is porous. It has tiny holes in it, so tiny you can't see them. Air can get into those holes and reach the inside of the charcoal. Coal is harder than charcoal and not as porous. It is even harder to start coal burning than charcoal. Starting a coal fire is quite difficult.

Still, every once in a while, coal *is* set on fire. Perhaps people start a campfire and a piece of coal gets kicked into it by accident. Or perhaps a piece of coal just happens to be lying on the ground where the fire is built.

Later on, when the fire goes out, someone might notice that in among the ashes is a glowing black stone. It keeps on glowing and it stays hot even when everything else has cooled down. The black stone is on fire, because if a twig or a blade of grass is put on the glowing part, it will begin to blaze.

It may have happened many times and the people who saw it may have thought, "Isn't that funny?" and then forgotten about it. Eventually, though, someone must have started looking for those black stones that burned. After all, they burn slowly and give off heat and it would be a lot easier to pick up these black stones than to cut down trees and chop them into firewood.

The first place where people started to burn coal deliberately seems to have been in China about a thousand years ago. (In those days China was the most advanced nation in the world.)

MARCO POLO

People in Europe knew nothing about what was going on in China at that time. In 1275, however, a young Italian named Marco Polo was taken by his family all across Asia to China, which was then the center of a huge empire.

Marco Polo stayed for many years and was astonished to see in how many ways China was larger, wealthier, and more civilized than Europe was. He finally got back to Italy in 1295, and three years later he wrote a book about his experiences in China. In this book he said, among other things, that the Chinese burned black stones as fuel.

Marco Polo's book was a best-seller, and many educated people read it all over Europe. They found it hard to believe some of what Marco Polo said, although we now know he was very truthful. Some of his readers must have wondered about the black stones, though. Perhaps someone who had seen black stones smoldering in a campfire decided to try to find some more.

During the next few hundred years, Europeans began to burn coal they found lying about. In the Netherlands Europeans got the idea for the first time of digging in the ground to find more coal.

This wasn't a very unusual thought. People dug in the ground to find ores from which metals could be produced and they dug for precious jewels. If there were inflammable black stones on the ground, might there not be more underground?

The people who dug in the Netherlands *did* find such underground coal.

People from the Netherlands traded a great

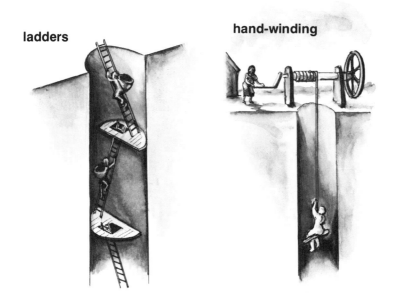

ladders

hand-winding

EARLY METHODS OF ENTERING A COAL MINE

deal with people from England, since the two lands faced each other across the North Sea.

The English noticed the way in which the Netherlanders were burning black stones, and there were some who knew that similar black stones could be found in England.

This was important to England, which together with Wales included about three fifths of the island of Great Britain. (The rest made up the nation of Scotland.) The English people needed wood for its furnaces, houses, and ships, but by 1600, much of the native forests were gone.

England could get wood from overseas, but that was risky.

England depended for its safety upon the ships it was building. The nations of Spain, France, and Austria all had larger populations than England and all had stronger armies. The only reason these other nations could not invade England and take it over was the fact that England was on an island. Invaders would have had to cross the sea and English ships managed to keep them away.

If England had to depend on wood from outside the island, what would happen if its enemies were to cut off its wood supplies? The English navy would suffer and dwindle and England might be lost.

England had to keep its own supplies of wood, which meant it had to use it very sparingly. If the English could find some other fuel to burn for heat and light, a lot of wood could be saved for the ships.

How about coal, then?

When the English started looking for coal, they found quite a lot in the north of the country. People dug for it and carried the coal by wagons to a town called Newcastle, which is on the North Sea coast of England. From there ships brought the coal to London.

Coal was just what England needed, and more and more of it was dug up in northern England and was shipped out of Newcastle.

By 1660 England was producing 2 million tons of coal each year. This was more than 80 percent of all the coal that was being produced in the world.

NEWCASTLE, ENGLAND, IN THE 1800s

At first, the coal that was brought to London was used chiefly as fuel. It was used to cook food and to warm the houses in winter. It was bituminous coal, and the content of tar and pitch caused it to burn with a smoky, smelly flame. Soot covered the city, which became dirtier than it used to be.

Still, the government encouraged this because it seemed better than burning scarce wood.

Wood still had to be burned in large quantities, however, to produce charcoal, for only charcoal could be used to obtain iron from its ores. Iron

43

had a million uses, and the English warships weren't going to be much good unless they had cannon made out of strong iron.

Furnaces for making iron had to be built in the forests, where there were lots of trees about. Such forests were far from where the iron was needed, and besides the forests were shrinking. The wood was being used up.

Could coal be used instead? Coal came in small pieces that could be carried anywhere so that the furnaces could be built near where the iron was needed.

The trouble was that bituminous coal wouldn't work. It didn't burn with a high enough temperature.

In 1603, however, an Englishman, Hugh Platt, discovered that if bituminous coal was heated in such a way that oxygen couldn't get at it, the tarry, pitchy material could be driven off and burned. Left behind was a material that was called *coke*. Coke was just carbon and was very much like charcoal. Coke burned with a high enough temperature to combine with iron ore and form iron.

At first, the coke wasn't of good enough quality to use. It took a long time to learn how to prepare good coke and how to use it to make iron. It wasn't till 1709 that an Englishman, Abraham Darby, began to use coke on a large scale for iron making.

This meant that, thanks to coal and coke, England's wood supply was saved.

Meanwhile, other Englishmen were trying to find some way of improving mines out of which coal and iron ore were obtained.

POLLUTED LONDON IN THE 1890s

In such mines water would collect at the bottom. This water had to be bailed out or pumped out if the mines were going to be useful. A lot of people had to do that bailing or work the pumps.

Could steam be made to do it instead? If you boil water in a kettle, the steam comes rushing out the spout. Perhaps such a current of steam could be used to blow the water out of the mine.

Or perhaps the steam could be allowed to fill a tube. Then, if the tube were cooled, the steam would turn back into water and there would be nothing in the tube, not even air. The tube would

**JAMES WATT WORKING ON A MODEL
OF NEWCOMEN'S ENGINE**

contain a *vacuum*. If the tube were stuck into water at the far end the water would rise to fill the vacuum. In this way, the water could be sucked out of the mine.

A "steam engine" for getting water out of mines was built in 1698 by an Englishman named Thomas Savery. It used steam under high pressure, and that meant there could be an explosion which could kill people.

In 1725 another Englishman, Thomas Newcomen who had been a partner of Savery, worked out a kind of steam engine that used steam at low pressure. This did the job and was much safer. By 1778 there were more than seventy Newcomen steam engines working in the mines of the single English province of Cornwall.

The trouble with any steam engine, however, is that first you must have steam to do the work. To get the steam you must boil water. To boil the water, you must burn fuel.

It took a great deal of fuel to produce the steam needed to do the work. Only one two-hundredths of all the heat produced by the burning fuel did the pumping. All the rest of the heat just warmed the metal out of which the engine was constructed and the air all about. This was a terrible waste of fuel.

In 1765 a Scottish engineer, James Watt, designed an improved steam engine that used six times as much of the heat of the burning fuel as the Newcomen engine did. (It was six times as "efficient.")

Watt continued to improve his design and his steam engine completely replaced the Newcomen steam engine. By 1800 there were about five

JAMES WATT
(1736–1819)

hundred Watt steam engines working in England.

What's more, Watt worked out ways by which the steam engine could be used to push and pull a piston that could make a wheel turn. This meant that steam engines didn't have to be used only for pumps. They could run machinery of all kinds faster and longer than human beings could.

In particular, steam engines were built that would run machines that would spin and weave threads mechanically. Cotton cloth could be made very cheaply in this way. (This was the beginning of what was called the *Industrial Revolution*.)

England had by then combined with Scotland to form the nation of Great Britain. Great Britain came to be a nation of factories that produced cotton cloth for all the world. It used the money it received to buy such materials as raw cotton. Since Great Britain sold the finished products for far more than the material cost, it quickly grew to be the richest and most powerful nation in the world.

Of course, even Watt's steam engine wasted about 90 percent or more of the heat of the burning fuel. If it had to use wood, Great Britain's forests would quickly have disappeared and its Industrial Revolution would have petered out.

However, the steam engines used coal, of which Great Britain had plenty. It was coal that made the Industrial Revolution possible.

Nor were steam engines used only in mines and in factories. Steam engines on ships could be used to make paddle wheels turn. The ships could then be made to move even against wind and current.

ROBERT FULTON
(1765–1815)

Such a steamship was first built by an American named John Fitch in 1787. He couldn't make his steamships earn money, however. In 1807, another American, Robert Fulton, built a steamship that proved to be a success. Little by little all trading ships became steamships.

On land, steam engines could be used to turn wheels. A steam engine built on a strong wagon could make it move without any horse pulling it, provided it were placed on smooth iron rails so there would be no unevenness to stop it. Such a wagon was a *locomotive* (from Latin words meaning "self-moving"). A locomotive could even pull a train of cars behind it that could carry people or freight. In this way, a "train" could travel along a "railroad."

The first successful locomotive was built as early as 1814 by a British engineer, George Stephenson.

The Industrial Revolution was changing the world very quickly, and it did so because of the burning coal that turned water into steam.

The Present
5 and Future
of Coal

GREAT BRITAIN was the first nation to go through the Industrial Revolution. The United States and Germany became industrialized toward the end of the 1800s, and this was made possible because both the United States and Germany had large supplies of coal within their borders.

In the 1900s Russia began to industrialize itself, too. It also had a great deal of coal. In fact, Russia (or the Soviet Union, as it is now called) produces more coal today than any other nation.

As more and more machinery came into use over more and more of the world, more and more coal was dug up and burned. Right now something like 3,000,000,000 (three billion) tons of coal are dug up and burned every year.

This wasn't entirely good. For one thing, burning coal produced soot and smoke. The large cities of industrial nations became dirtier and dirtier.

Anthracite coal burned with far less smoke than bituminous coal, but anthracite coal isn't as com-

OIL TRANSPORT

mon. Bituminous coal can be treated to make it burn in a cleaner fashion, but that would make it more expensive.

Then, too, people had to dig deeper and deeper to get coal and doing so was dangerous. People died in explosions, in cave-ins, and from diseases of the lungs produced by the coal dust. It was hard to carry the heavy coal from the mines to where it was used, and coal was a heavy material to shovel into furnaces and hard to set on fire.

In the latter 1800s people began to turn to liquid oil for fuel. It was not so much the oils that

could be obtained from plants or animals that were used. There just wasn't enough of that. It became possible, however, to pump oil out of the ground. This was called *petroleum* or just *oil*. Petroleum could be purified into different varieties such as *kerosene* and *gasoline* (see *How Did We Find Out About Oil?* Walker, 1980).

There were a great many advantages to oil. It didn't have to be mined. A deep hole could be drilled into the ground and (if the hole was in the right place) the oil could be pumped up. Nobody had to go underground to get it.

What's more, oil is easy to transfer from place to place. Instead of having to be carried overland on long freight trains as coal has to be, oil can be pumped through long pipelines that sometimes stretch for thousands of miles. At the ocean's edge it can be pumped into oil tankers that can take it across the ocean.

Oil is much easier to handle at the furnaces, too. It can be pumped into the furnaces at just the right rate. It can be set to burning very easily and made to stop burning just as easily. It leaves no ash.

Some of the substances obtained from oil can be used to run automobiles, buses, trucks, ships, airplanes. It could be used to heat houses, run steam engines, make electric generators work.

About the only thing oil couldn't do that coal could was to turn iron ore into iron. Making iron and steel was still the job of coal.

Little by little, as the 1900s wore on, oil began to gain on coal as a fuel. After 1950 oil became the most important fuel in the world. From 1950 on, the whole world was industrializing faster than ever and it was doing it by means of oil.

There was a catch to oil just the same. There was far less oil in the ground than there was coal. In the 1970s it began to look as though the oil would not last very much longer. It looked as though there just wouldn't be enough oil in the 1980s to produce the energy required.

What's more, most of the oil was in the Middle East, which is an unstable part of the world. Beginning in 1973 the oil-producing nations began to raise the price of oil rapidly.

56

COAL FIELDS IN THE UNITED STATES AND CANADA

What is the world going to do?

It depends on machinery for almost everything. For instance, machinery helps the world produce much more food than it could in the days before the Industrial Revolution. The world's population is now nearly five times as great as it was when Watt first produced his steam engine.

If oil ran out and machinery shut down, most of the world's population might starve.

One thing we can do to keep that from happening is to go back to coal. There is still plenty of coal in the ground. There is enough to last the world for several centuries.

We have made advances and can continue to make more advances. People have learned to make use of machinery to mine coal so that it takes fewer miners these days to mine more coal and to do it more safely.

Then, too, it might not be necessary to carry all the coal long distances to where they can be used. Ways will probably be worked out to combine the coal with hydrogen right at the mine. This will turn coal into a liquid fuel that will be every bit as convenient as oil.

Even so, coal continues to have disadvantages. Digging it out of the ground destroys the soil and pollutes the waters round about. Burning it creates smoke and soot and air pollution. Of course, the ground can be restored and the coal can be cleaned up to lower the pollution, but that takes a great deal of time and effort. It would make coal far more expensive to use.

The supply of coal is limited. There is far more coal than oil, but the coal will be used up someday, too. What then?

In fact, it may not even be safe to burn all the coal we can dig out of the ground. You see, coal is mostly carbon and when the carbon combines with oxygen, it forms a gas called *carbon dioxide*.

Carbon dioxide is not a dangerous substance. It can be even called a very useful substance.

There is a little in the air at all times, about 3.5 pounds in every 10,000 pounds of air. It is very important that that little bit should be there. Green plants live on the carbon dioxide. They use the energy of sunlight to combine the carbon dioxide with water and minerals to form the plant tissue so it can grow.

The green plants might run out of carbon dioxide. When animals breathe, however, they produce carbon dioxide. That restores what the plants use up.

If there were no carbon dioxide in the air, there would be no plants, no animals, no human beings.

In that case, is it a good thing to burn coal and produce more carbon dioxide. Will this help grow still more plants?

Not entirely, it seems. We are producing carbon dioxide at a greater rate than the plant world can make use of it. Ever since 1900, the carbon dioxide has been piling up in the air. It is now 3.5 parts in 10,000; in a few years it will be 4 parts in 10,000. That still isn't much. It won't interfere with our breathing, for instance. But there is a catch.

Sunlight reaching the earth passes through the atmosphere and delivers heat to the earth. At night, the earth radiates the heat it has gained from the sun back into space. In this way, the average temperature of the earth remains unchanged. What heat the earth gains during the day, it loses at night.

The heat the earth radiates away is in the form of *infrared*. This is made up of waves like those of light, but longer. The long waves of infrared are absorbed by carbon dioxide in the air (though the short waves of ordinary light are not). The fact that infrared is trapped means that the earth is a little warmer than it would be if there were no carbon dioxide in the air.

COAL MINER

As the carbon dioxide content of the air goes up, even slightly, more of the infrared is absorbed and the average temperature of the earth goes up also. It doesn't go up much, but it goes up enough, possibly, to change the climate of the earth.

Right now scientists are trying to figure out what changes in the climate might take place and how damaging it will be. If we have to stop producing carbon dioxide, we would have to switch to other ways of running machinery than by burning coal.

We might have to burn hydrogen. We might have to use wind and running water to produce electricity. We might have to use nuclear energy. We might have to use sunlight.

It can all be done, but it will take time and effort.

Still, even if the time comes when we stop using coal, we should always remember that it was coal that made the Industrial Revolution possible. It is coal that made the modern world come into existence with its machinery and its new inventions.

Coal made it possible to feed more people and make them more comfortable.

And it did it all only in the last three hundred years. Before that, coal was only a black stone that hardly anyone even knew existed.

STRIP MINE

Index